Super Science

Materials

Richard Robinson

QED Publishing

First published in the UK in 2007 by
QED Publishing
A Quarto Group company
226 City Road
London EC1V 2TT
www.qed-publishing.co.uk

A catalogue record for this book is available from the British Library.

ISBN 978 1 84538 664 1

Written by Richard Robinson
Edited by Anna Claybourne
Designed by Balley Design Ltd
Consultant Terry Jennings

Publisher Steve Evans
Creative Director Zeta Davies
Senior Editor Hannah Ray

Printed and bound in China

Picture credits
Key: T = top, B = bottom, C = centre, L = left, R = right, FC = front cover

Corbis: p13 Michael S. Yamashita; p14 Marcelo Sayao/epa; p19 Guillaume Bonn; p20 Werner Nosko/epa; p23 Bernard Annebicque/Sygma; p25 Bettmann.

Science Photo Library: p4 NASA/ESA/STSCI/J. Hester & P. Scowen, ASU; p5 Erich Schrempp; p8 p10 Dr. Juerg Alean; p18 Sheila Terry; p22.

Words in **bold** can be found in the Glossary on page 31.

Contents

What are materials?

The word 'materials' means stuff. Materials are all the different types of stuff that make up everything around us, and everything in the Universe. Stars, planets, rocks, trees, plastic, sand, water, food, clothes and animals are all made of materials. We are made of materials, too.

Where did materials come from?

Scientists have a theory to explain where materials came from, and it goes like this. Around 13.7 billion (13 700 000 000) years ago, there was a sudden explosion, known as the Big Bang. At this moment, time, space and matter (or materials) came into being. Before the Big Bang, nothing existed at all.

To begin with, there were just two types of materials, called hydrogen and helium. Gravity pulled together hydrogen and helium gas to form stars. The stars exploded in smaller 'bangs', forming vast clouds of dust and creating the planets and all the other materials which we find around us today. This means that everything around you is made of materials that were formed in the centres of stars. And that includes you – so you are made of stardust!

Right: The Eagle nebula is a 'star nursery' in space. It is a vast cloud of hydrogen and helium gas and dust, which gravity is pulling together. This will eventually create new stars.

Above: No one knows what the Big Bang really looked like, as no one was there to see it – but this is an artist's impression.

We weren't there to see the Big Bang, so what makes scientists think it happened? See page 30.

Right: Everything in your bedroom is made of materials.

Atoms and molecules

All materials are made up of very, very small building blocks called **atoms**. Billions and billions of tiny atoms packed together make up every object – such as the Sun and stars, rocks, trees, this book, your shoes, and your body.

How small is an atom?

Atoms are so small that it's impossible to see one atom on its own, except through a very powerful microscope. A single page in this book is about a million atoms thick. The air around you contains billions of atoms in every cubic centimetre, but you can't see them.

Right: Everything you can touch and hold is made of atoms.

Look inside an atom

Atoms themselves are made up of even tinier bits. In the middle of an atom is a ball of particles called the **nucleus**. Other, extremely small particles called **electrons** zoom around the nucleus in a cloud.

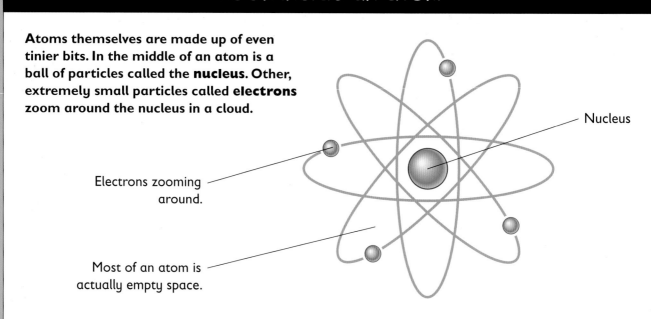

Nucleus

Electrons zooming around.

Most of an atom is actually empty space.

Atoms and molecules

Atoms can join together (bond) to make bigger building blocks called molecules. For example, one oxygen atom can bond with two hydrogen atoms to make a water molecule. Water in the sea, your bath or a drink is made up of millions of water molecules like this. Atoms can combine in all kinds of patterns, to make millions of different kinds of molecules.

Whenever you touch something, such as your desk, a drink or your own face, you are feeling masses and masses of atoms, held tightly together to form molecules. If you could zoom closer and closer, until you could see them, you'd see that everything, everywhere, is made of them.

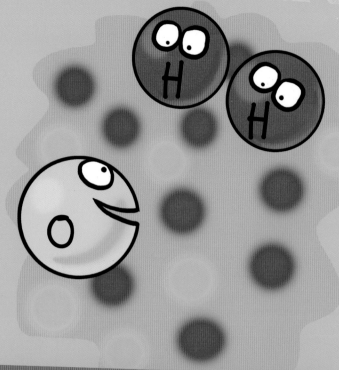

Right: All the water around us is made up of tiny water molecules. Each water molecule is made of two hydrogen atoms and one oxygen atom.

Making molecules

Scientists give atoms and **molecules** names made of letters and numbers. The name for water is H_2O. This means it has two hydrogen atoms and one oxygen atom.

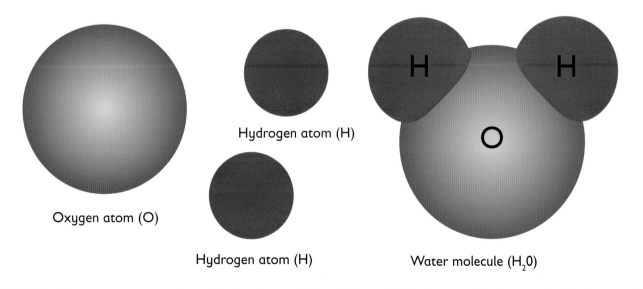

Oxygen atom (O)

Hydrogen atom (H)

Hydrogen atom (H)

Water molecule (H_2O)

Elements

Elements are the most basic, pure types of materials. All other materials are made of different combinations of elements.

What is an element?

An element is a substance that is made of just one type of atom. For example, gold is an element, as it contains only gold atoms. There are 92 types of atoms that occur naturally on Earth, so there are 92 natural elements. (Because scientists know how to build atoms, they have been able to construct some other elements, too. But these are very unstable, and only survive for a few trillionths of a second.)

Six elements make up 99% of the human body. Can you guess which ones? See page 30.

Everyday elements

Although some elements sound mysterious and strange, many are normal substances that we encounter every day.

Iron is an element – it's used to make all kinds of things, from pots and pans to railway tracks. Mercury, found inside some thermometers, is also an element. So are oxygen, which living things breathe in from the air, and helium, which is used to fill floating party balloons.

Above: These rings are made of gold, one of the Earth's natural elements.

Above: In 1869, Russian scientist Dmitri Mendeleev worked out the different qualities of all the elements and arranged them into a handy table. It's known as the **Periodic Table of the Elements**.

Spot the elements

These are the names of all the naturally occurring elements.
How many have you heard of? (You might recognize up to 30).

Hydrogen	Calcium	Rubidium	Barium	Rhenium
Helium	Scandium	Strontium	Lanthanum	Osmium
Lithium	Titanium	Yttrium	Cerium	Iridium
Beryllium	Vanadium	Zirconium	Praseodymium	Platinum
Boron	Chromium	Niobium	Neodymium	Gold
Carbon	Manganese	Molybdenum	Promethium	Mercury
Nitrogen	Iron	Technetium	Samarium	Thallium
Oxygen	Cobalt	Ruthenium	Europium	Lead
Fluorine	Nickel	Rhodium	Gadolinium	Bismuth
Neon	Copper	Palladium	Terbium	Polonium
Sodium	Zinc	Silver	Dysprosium	Astatine
Magnesium	Gallium	Cadmium	Holmium	Radon
Aluminium	Germanium	Indium	Erbium	Francium
Silicon	Arsenic	Tin	Thulium	Radium
Phosphorus	Selenium	Antimony	Ytterbium	Actinium
Sulphur	Bromine	Tellurium	Lutetium	Thorium
Chlorine	Krypton	Iodine	Hafnium	Protactinium
Argon		Xenon	Tantalum	Uranium
Potassium		Caesium	Tungsten	

Solids, liquids and gases

Materials such as water, metals and rock can exist in three different forms – solids, liquids and gases. These three forms are called the states of matter. A material can change from one state to another, too. For example, when butter melts, it's changing from a solid to a liquid.

Look around

Solids and liquids are easy to spot – look around now and you might see solid furniture, cars or trees, and liquid rain or drinks. Gases, on the other hand, are usually invisible, so we can easily forget about them. (There are gases floating around your head right now, including oxygen, nitrogen and carbon dioxide. You wouldn't know that by looking, would you?)

One substance that is familiar to us in all three states is called water when it is a liquid, known as ice when it is a solid, and is called water vapour when it's a gas. Many materials exist mainly in one state, and we rarely see them in another state. For instance, rocks are usually solid, but inside a volcano they can get so hot that they melt and pour out as a liquid called lava.

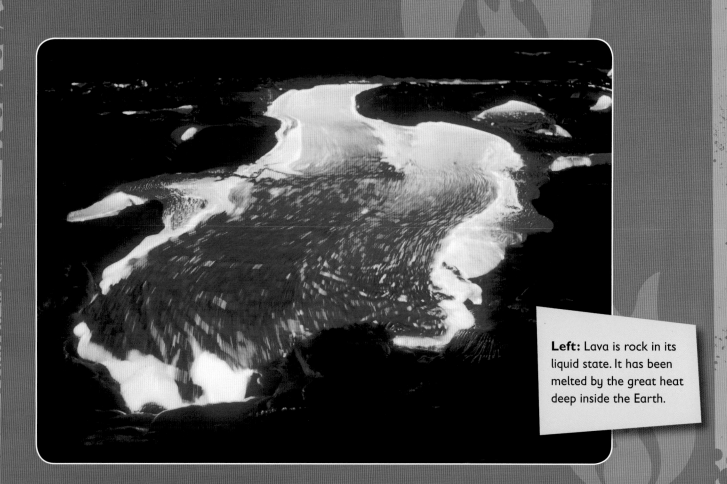

Left: Lava is rock in its liquid state. It has been melted by the great heat deep inside the Earth.

The states of matter

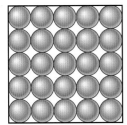

In a solid, atoms and molecules sit close together.

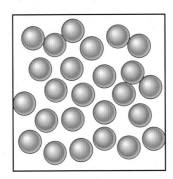

In a liquid, they flow around more freely.

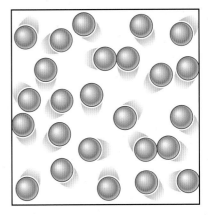

In a gas, they shoot around and are spaced wide apart.

Changes of state

How can a material change between solid, liquid and gas? It's all to do with temperature. The atoms and molecules in a material are always jiggling about and bouncing into each other. When they are cold, they jiggle less, and they can sit close together, forming a solid. Heat makes them move more, and push each other apart to melt, or become a liquid. Even more heating jiggles them so hard they fly apart, forming a gas.

As a material gets warmer, its molecules move further apart, so it takes up more space. Most materials expand, or get bigger, as they get hotter.

Right: There are solids, liquids and gases all around you.

Changing materials

Besides changing between solid, liquid and gas, there are other changes that can happen to materials. Sometimes, once a change has happened, it can't be changed back.

Heating and cooking

Instead of melting, some materials behave very differently when they are heated. When soft, slippery clay is heated to a high temperature, it becomes extremely hard – perfect for making pots and plates. Our ancestors made this discovery about 12 500 years ago.

It was also found that mealtimes got better when fire was used. Our ancestors ate food raw until they found that cooking food made it easier to digest. Mixing ingredients and cooking them together to make them tasty came soon after.

You probably wouldn't want to eat raw eggs, flour, sugar, butter, salt and baking powder. But if you mix and heat them, they become something completely different – a delicious cake.

Salt evaporation

If you mix salt with water, the salt will dissolve and disappear into the water. To get the salt back, pour the salty water onto a saucer and leave it for a day or two. The water will evaporate (turn into vapour, or gas), leaving the salt behind on the saucer.

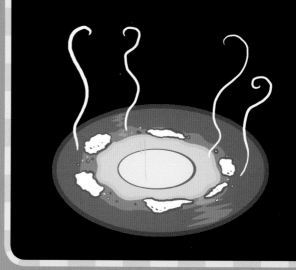

Left: Cake isn't just a mixture of ingredients. Cooking changes the molecules to make a new material.

Above: Firing turns clay into hard pottery. To fire their pots, potters use a very hot oven called a kiln.

Reversible and irreversible changes

Some changes are irreversible – once done, they cannot be undone. A pot cannot be ground down and turned back into wet clay, and a cake cannot be unbaked. The molecules that made up the ingredients have rearranged themselves to form new molecules – and new materials.

Some changes are reversible, which means they can be changed back. For example, you can mix salt and water to make salty water, and then turn it back into salt and water vapour. (see experiment box opposite).

Can you think of some other changes that cooking causes? What happens to an egg, a potato or a slice of bread when they are heated? See page 30.

Animal, vegetable or mineral?

So materials can be divided into solids, liquids and gases. Another way to divide them up is to ask where they came from — an animal, a vegetable (meaning a plant), or a mineral, such as a rock.

Which is which?

All materials can be classed as animal, vegetable or mineral, or a combination of them. Look at the list below, and see if you can **classify** each material into animal, vegetable or mineral. The answers are opposite.

Newspaper

Washbasin

Aspirin

Glass bottle

Bath

Pepper

Plastic bottle

Toothpaste

Salt

Clothes

Bicycle tyre

Pencil lead

Can you think of any everyday objects that include animal, vegetable and mineral parts? See page 30.

Left: Every day, thousands of trees are turned into daily newspapers.

Paper-making

Tear up some old newspaper into tiny pieces, and soak them in a little water overnight. Ask an adult to liquidize the mushy paper in a food blender. Spread the grey goo out between two teatowels or pieces of old cloth, using a rolling pin to roll it flat. Leave it to dry in a warm place. Peel off the cloth, and there is a sheet of your own paper!

Answers

Newspaper: vegetable.
Paper is made from trees. One fir tree will produce a pile of newspapers 1.2m high.

Glass bottle: mineral.
Made from silicon dioxide, the chemical name for sand.

Plastic bottle: animal and vegetable.
Plastic comes from oil, which is the remains of tiny sea plants and animals

Clothes: to find the answer, look at the label. Woollen clothes are made from sheep's wool (animal), while thinner clothes are often made from cotton plants (vegetable).

Washbasin: mineral.
Made from clay, a kind of rock.

Bath: mineral or vegetable.
Made from metal, plastic or fibreglass, which is made from glass.

Toothpaste: mineral or vegetable.
A mixture of chalk, glycerin (from plants) and antiseptic chemicals (mineral), plus seaweed juice and mint (both vegetable).

Bicycle tyre: vegetable.
Made from rubber, which is made from the sap of the rubber tree.

Above: Table salt is actually a mineral called sodium chloride. It's one of the few minerals we eat.

Aspirin: mineral.
Today made from a variety of minerals, although it was first discovered in the bark of the willow tree.

Pepper: vegetable.
Made from the ground-up seeds of the pepper plant.

Salt: mineral.
Made from sea water or dug up from the ground.

Pencil lead: mineral and vegetable.
Made from a mixture of clay (mineral) and graphite, a kind of coal made from compressed plants (vegetable).

Water – a special material

For life on Earth, water is an essential material. And although it is very common, it is quite strange. It behaves differently from most other materials.

Shaping the land

Water doesn't shrink when it freezes into ice, or expand (get bigger) when it melts back into water. In fact, water expands a little when it freezes – and this is why the planet looks the way it does.

Water seeps into cracks in the rocks of mountains and cliffs. When the weather gets cold, the water freezes and expands. As it expands, it pushes the rocks outwards, splitting them, so that bits fall off. This makes rocks break apart and crumble away, or erode – shaping mountains, cliffs and coasts.

When scientists look for life on other planets, they search for water. Can you think of anything else that might indicate the existence of life? See page 30.

Right: Rain brings fresh water onto the land, allowing plants and animals, including humans, to live in almost all parts of the world.

How water can break a rock in two...

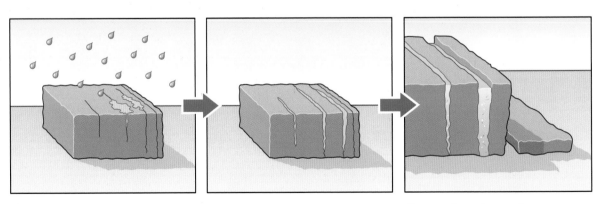

Water seeps into cracks in the rock.

In cold weather, the water freezes and expands.

This pushes the rock apart, splitting it into pieces.

Water for life

Water flows around our planet in a huge cycle called the water cycle. The Sun heats water in the seas. This makes the water molecules move faster and turn from a liquid to a gas (water vapour), which floats into the air. High in the sky, the cool air turns the water vapour back to liquid water in the form of tiny droplets, which form clouds. When the droplets are big enough, they fall as rain, supplying plants and animals with water. The water flows into rivers and finally out to sea again.

But there is an even bigger reason why water is important: it is the basis of all life on Earth. You, and all the animals and plants on our planet, would not exist if it were not for water. In fact your body is about 70 per cent water.

The water cycle in your kitchen

Ask an adult to put the kettle on and hold a metal ladle (with the handle wrapped in a tea towel) upside-down near the spout as it boils. First, the heat turns the water into invisible water vapour, just above the spout. Then water droplets form, which you can see as clouds of steam. When the steam touches the cold ladle, it collects as drops of liquid water, similar to raindrops. You can catch the drops on a plate.

Left: Earth has far more water than the other planets we know of. Oceans and seas cover 70 per cent of its surface and make the planet look blue from space.

The quest for gold

For centuries, one material has appealed to humans more than any other – a yellow metal element, called gold. It has always been valuable, and in the past it was used as money.

Recipes for gold

The main force behind all modern chemistry was the ancient craft of alchemy. Hundreds of years ago, alchemists cooked up all sorts of weird materials together, trying to make gold. They tried thousands of ways and spent millions of pounds in their efforts. But it never happened. The only way to get gold is to dig for it in the ground. Many alchemists actually made themselves very poor in their effort to make themselves rich.

In the process, the alchemists made some extraordinary discoveries. 'Alchemy' turned into 'al-chemistry', and at last into 'chemistry', the science of how atoms and elements bond together and change.

What other uses does gold have in the modern world? Can you think of any? See page 30.

Right: The German alchemist Hennig Brand saw that **urine** had a golden colour, and this gave him an idea. He would extract gold from urine! He boiled down buckets of urine over months until he was left with a solid lump. Of course, it wasn't gold. But the brown soapy block did do something magical – it glowed in the dark! Brand had accidentally discovered another useful chemical, phosphorus.

Above: Gold is found in the ground as nuggets, or as narrow strips or seams in rock.

Pure gold

What made gold especially attractive and desirable to our ancestors was its purity. It doesn't combine with other elements, and it doesn't tarnish or rust, but stays shiny. However, apart from jewellery, gold has very few uses today – mainly because of that same unwillingness to combine with other elements. To make new materials, chemists like to use more 'reactive' elements, which combine easily with each other. So gold is now mostly used just for decoration and ornament.

Right: Since gold doesn't react with other materials, gold objects last and keep their value. Gold jewellery buried in the ground will last for thousands of years.

Material ages

Throughout history, people have learned more and more about how to use the materials around them to make useful things. The materials they used give their name to the different 'ages', or time **periods**, of early human history.

The Stone Age

The earliest people used the materials around them in their simplest forms – things like wood, boulders and bones. Very early human history, from about two million years ago to five thousand years ago, is known as the **Stone Age**, because most tools were made from stone. People discovered a kind of stone called flint, which is very hard and smooth, almost like glass. When struck with another rock, it breaks into sharp fragments that make good blades and arrowheads. Flint tools enabled humans to hunt animals, cut them up for meals, and make their skins into clothes and their bones into tools and ornaments.

Below: Ötzi is a preserved 5300-year-old man found in a **glacier** in Europe in 1991. In his possession, among other tools, were knives made of flint and wood.

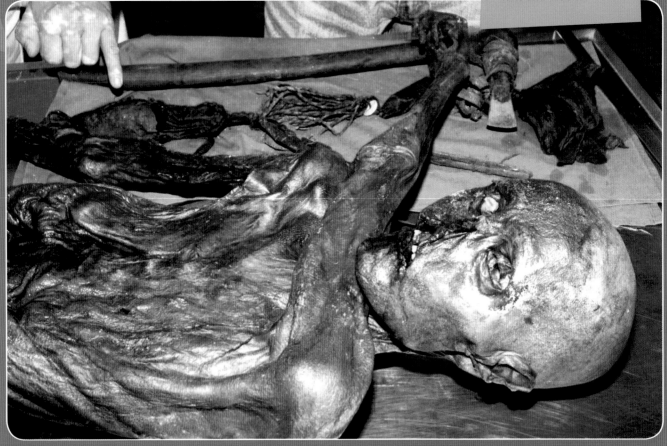

Stone tools

Archaeologists, who study evidence of how people used to live, have found thousands of flint tools from the Stone Age.

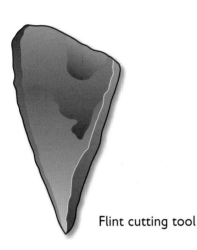

Flint cutting tool

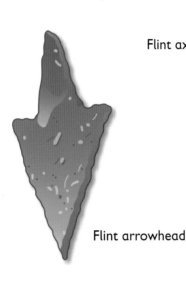

Flint arrowhead

Flint axehead

The Bronze Age and the Iron Age

Stone-Age people eventually noticed that some types of stones would melt in the fire, turning into strange, shiny substances that were incredibly useful, being strong and bendable. These were metals, and from them they made bronze (a mixture of two types of metal, copper and tin). They could throw away their stone tools now.

The **Bronze Age** lasted for 2000 years, until people found something even better for making tools with – iron. Now came the **Iron Age**.

The Stone, Bronze and Iron Ages are names given to earlier times. What would you call the present age? See page 30.

Using materials

Just like Stone-Age and Bronze-Age people, we use the materials around us to make useful things. Yet we have come a long way since **prehistoric** times.

Material science

As well as identifying all 92 naturally occurring elements, and learning to create new ones, scientists combine different elements to make new materials. They can also recreate some natural materials in the lab. Materials that have been designed and made in laboratories are known as synthetic materials.

However, we still use ancient, natural materials, even in the most modern inventions. For instance, computers contain a little bit of the Stone Age, as their processing chips are made using silicon – the main ingredient of flint.

Right: The rare metal element tungsten can be used to make hip replacements, because it is very strong, and doesn't rust. But scientists are also developing new materials that could be used to make artificial hips, including new types of plastics and ceramics.

There is one problem with turning platinum into gold. Can you guess what it is?
See page 31.

Above: Kevlar is used for protective clothing. It is made from strands of a special type of plastic. It is strong, light, rust-free and fire-resistant.

Making plastic

Mix half a cup of warm milk with a splash of vinegar. Pour the mixture into a handkerchief and squeeze the liquid out. The solid that's left will dry really hard – and is a form of plastic!

New materials

Today, scientists are busy inventing thousands of brand-new materials with ever more advanced uses. For example, medical scientists make new drugs for treating diseases by putting different types of atoms together to design new molecules. Plastics scientists experiment with different molecular structures to build brilliant new materials for all kinds of modern uses. In 1966, US scientist Stephanie Kwolek invented the new, ultra-strong material Kevlar, which is used in bulletproof vests, sports gear and spacecraft.

We now know so much about how materials work that we can even make gold! By altering the basic structure of a lump of another metal, platinum, we can change it into gold.

The right material

Different materials have many different properties – that is, the things they can do and the way they behave. For example, wood floats, copper is flexible, gold doesn't rust and steel is very strong. So when we use materials, we have to choose the right one for the job we need it to do.

New and improved

Humans learned early on that some materials are better than others for certain jobs. They used flint to make tools because when it chips or breaks, it has very sharp edges. (One scientist investigating a Stone-Age cave dwelling cut himself on a million-year-old axe!)

However, flint breaks easily, so when long-lasting, flexible bronze came along, it did the job even better. But bronze was very heavy, so it wasn't much use for making weapons or armour. Iron is much lighter, and iron blades keep sharper for longer – so iron took over from bronze. Adding carbon to iron made steel, which is even stronger. Adding a little chromium makes 'stainless steel', which does not rust. As you can see, the search for better and better materials for different jobs drives us to invent more new materials.

We don't make chairs with rubber legs, or umbrellas out of newspaper, because they wouldn't work. Choosing the right material is all about getting objects to work as well as they can.

Spaghetti towers

Dry spaghetti is very brittle, but it's surprisingly strong lengthways. You can use it to make amazingly tall towers. Use pieces of marshmallow to join the spaghetti strands together.

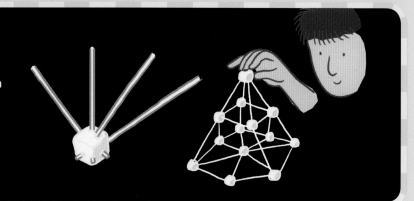

The wrong materials

All around you are objects made of suitable materials. You can see how suitable they are by imagining them being made of the wrong material. We don't make house bricks out of metal, and we don't make cranes out of stone. Can you think why? What about trousers made of wood, a spoon made of sponge, or a chocolate saucepan? Why wouldn't they work?

What's the most amazing material you can think of? See page 31.

Below: The airship Hindenberg, the biggest aircraft ever, exploded in 1937, killing 36 people. The disaster was partly down to a bad choice of materials. The airship was filled with hydrogen gas, and the outside was covered with aluminium paint. Both of these catch fire easily, making the ship very unsafe.

Materials into energy

As well as using materials to make useful things, we use them as fuel. Materials contain energy in their atoms and molecules, which we can **convert** into useful forms of energy such as movement, heat and electricity.

Plant power

Since discovering fire, humans have turned plants into useful heat energy, by using wood as firewood and burning it. More recently, we have discovered how to use oil and coal. These are made from ancient plants and animals squashed under layers of mud and rock. Oil can be used as fuel in cars and planes, making them move, or burned in power stations to generate electricity. (Oil has another important use too — it's used to make different kinds of plastics.)

Below: To collect all the oil we need to make energy, we have to drill into the ground or under the seabed to where the Earth's stores of oil are found. This oil rig is collecting oil from beneath the seabed in the North Sea.

How we use energy from prehistoric creatures

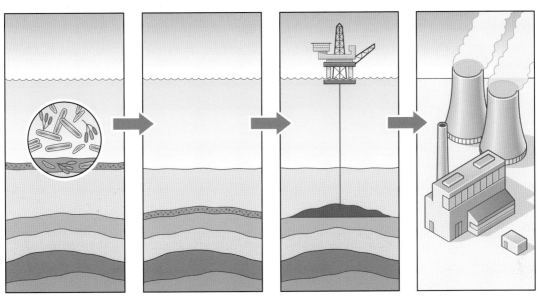

1) Millions of years ago, tiny plants and animals floated down to the seabed.

2) They were gradually squashed under layers of mud, and turned into oil.

3) We pump oil out of the ground and then pipe it to power stations.

4) Power stations burn oil and convert the heat energy into electricity.

Nuclear power

Besides burning, there is another way to harness energy from materials: nuclear power. To release this power, we have to split atoms apart and release the powerful energy stored within them. The task is not easy, the dangers are great and the left-overs are poisonous, but being able to use the fundamental building blocks of the universe to make energy is very exciting. Many countries now have nuclear power stations to provide some of their energy. Scientists are also trying to find ways of creating energy by fusing, or joining, atoms together instead of splitting them apart. This is the process that makes the Sun give out energy, in the form of heat and light.

Above: Power stations that burn coal or oil cause air pollution from waste gases. Nuclear power does not.

Using materials wisely

For a long time, we've been able to take whatever materials we wanted from the Earth to meet our needs. But if we're not careful, some of them will soon run out.

Making materials last

Many raw materials are becoming more difficult and more expensive to find. For example, copper, coal, oil and uranium, which is used in nuclear power stations, are all running low. Yet when we have finished with things we throw them away, assuming that we can always find more. This cannot go on for ever.

Instead, we need to learn the three Rs – Reduce, Re-use, and Recycle. This means that we must reduce the amount of stuff we use, re-use things rather than throw them away, and if we can't re-use, then recycle the material.

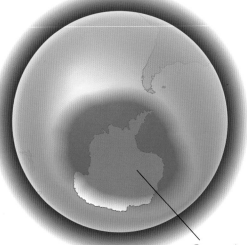

Right: The average person uses 133 plastic bags per year, then throws them away after one use. It's easy to save this waste of precious materials by using a re-usable cloth bag instead.

The ozone layer

CFCs (see opposite) have damaged the **ozone layer** around the North and South Poles. As ozone protects us from some of the harmful effects of sunlight, governments had to move fast to ban the ozone- destroying chemicals.

Gases from aerosols and fridges have damaged the ozone layer.

Ozone hole

Above: A landfill site is a place where we dump things we no longer want. While some of our rubbish rots away safely, the rest is made of materials that last for ages or, even worse, release poisonous chemicals into the ground.

Material safety

Materials science isn't just about making new things – we also need to make sure new materials are safe. In the 1920s, a new gas was invented, called chloro-fluoro-carbon (CFC). It was used in refrigerators, fire extinguishers and aerosol sprays. For many years it was thought to be safe.

In the 1980s, scientists discovered that CFC gas was floating high into the Earth's atmosphere, and starting to destroy a layer of gas called ozone, which protects us from the harsher effects of sunlight. CFC gases have been banned now, but this story is a warning to us all to be careful with new materials we create.

Does your household practise recycling? Have you noticed the difference between some plastics, which can be recycled, and others which cannot? See page 31.

Questions and answers

We weren't there to see the Big Bang, so what makes scientists think that it happened? (page 5)

There are several reasons. The main one is that measurements show that the Universe is expanding and stars are moving apart. So they must once have been much closer together.

Six elements make up 99% of the human body. Can you guess which ones? (page 8)

Oxygen: 65%. Carbon: 18%. Hydrogen: 10%. Nitrogen: 3%. Calcium: 2%. Phosphorus: 1%.

Can you think of some other changes that cooking causes? What happens to an egg, a potato or a slice of bread when they are heated? (page 13)

An egg changes from runny to hard, a potato changes from hard to soft, and a slice of bread goes dark and crispy.

Can you think of any everyday objects that include animal, vegetable and mineral parts? (page 14)

A shoe might have a leather upper (animal), a rubber sole (vegetable), and metal buckles or shoelace holes (mineral). A handbag could be made of leather (animal) with a cotton lining (vegetable), and a metal zip (mineral).

When scientists look for life on other planets, they search for water. Can you think of anything else that might indicate the existence of life? (page 16)

Strangely, the other thing scientists search for is methane (a gas found in animal dung). They reckon that if something is living it must be producing methane, which can be detected in the atmosphere even in tiny amounts.

What other uses does gold have in the modern world? Can you think of any? (page 18)

Gold is still used to make gold teeth, non-rusting connectors in electrical circuits, and gold medals, and as a coating on astronauts' helmet visors to protect against strong sunlight.

The Stone, Bronze and Iron Ages are names given to earlier times. What would you call the present age? (page 21)

The modern age is often called the 'Nuclear Age' because we can manipulate parts of atoms (the centre of an atom is called the nucleus) and make nuclear power. It could also be called the 'Plastic Age'. Do you think it will last as long as the Stone Age?

There is one problem with turning platinum into gold. Can you guess what it is? (page 22)

Platinum is more expensive than gold to begin with, so it's not worth turning it into gold.

What's the most amazing material you can think of? (page 25)

There are many amazing materials, including silicon – the main ingredient of sand and many types of rock – which is used in making computers.

Does your household practise recycling? Have you noticed the difference between some plastics, which can be recycled, and others which cannot? (page 29)

Most plastic can be recycled, some types of food wrapping cannot. Check with your local authority for details of what they will and will not accept for recycling.

Glossary

Atoms – The tiny particles that all matter is made up of. Atoms are very small. If people were as small as atoms, the whole population of the Earth could fit onto the full stop at the end of this sentence 200 times over.

Bronze Age – The period from about 5 500 to 3 500 years ago, when people developed the use of bronze (a mixture of copper and tin) to make tools and weapons.

Classify – To divide things into different groups or classes.

Convert – To change something from one thing into another.

Electron – A tiny part of an atom, which flies around the nucleus, or middle of the atom, at high speed.

Element – A pure substance made of just one type of atom.

Glacier – A slow-flowing river of ice formed from crushed snow. Glaciers are found in cold mountainous regions and near the poles.

Iron Age – The period from about 3 500 years ago to 2 000 years ago, when iron replaced bronze as the metal of choice for tools and weapons.

Molecule – A particle made up of atoms joined together. For instance, a molecule of water is made of atoms of hydrogen and oxygen.

Nucleus – The middle part of an atom, made up of a cluster of particles called protons and neutrons.

Ozone layer – A layer of ozone, a type of oxygen, found high in the Earth's atmosphere. It protects us from the Sun's most harmful rays.

Period – In history, period means a length of time. In chemistry, it means the horizontal row of elements in the Periodic Table.

Periodic Table of Elements – A chart which shows all the elements arranged according to their atomic number, beginning with the lightest (hydrogen).

Prehistoric – Relating to the time before historical records were written down.

Properties – The qualities of a substance, such as how it behaves, how much it weighs, its colour and so on.

Stone Age – The period from about 2 million years ago to 5 500 years ago, when people used stones to make their tools.

Urine – Liquid waste from the human body, otherwise known as pee or wee.

Index